The Opal Story

A GUIDEBOOK

Andrew & Damien Cody

Contents

Birth of the Opal

One of the Australian Aboriginal Dreamtime stories, as told by June Barker of Lightning Ridge.

In the Euraliah* country, the Dreamtime creator came down to earth in a giant rainbow. He gathered together all the tribes and said he would return when he thought they were wise enough to carry out his plan to have peace forever on earth.

On the stony ridges where the rainbow had rested there was a great area of rocks and pebbles. Next morning, when the sun rose and shone his light on that spot, the rocks and pebbles flashed and glittered in the sun with all the colours of the rainbow that had given them birth: red, orange, green, yellow, blue and violet. These were the first opals.

** Euraliah country is the traditional country of the Euraliah people, from Lightning Ridge to the Narran Lakes in northwestern New South Wales, Australia.*

The Opal in History:

"The most highly prized of all gemstones"

Opal derived its name from "Opalus" which meant "to see a change in colour". The Ancient Romans popularised opals more than two thousand years ago, when opals were found at Czernowitz, a mountainous region of Slovakia, which was at that time part of Hungary.

Pliny the Elder wrote the inaugural Natural History of the World, which was first published in 77AD. In this most important publication he wrote that opal was the most highly prized and valuable of all gemstones in the Empire.

Pliny the Elder. *Cleopatra and Marc Anthony.* *Coronation of c*

Queen Victoria. *Louis Comport Tiffany.*

nan Emperor. William Shakespeare. Napoleon and Josephine.

René Lalique. Australian opal pioneers E. Murphy, Tullie Wollaston, Charles Nettleton.

Famous People and the Opal

Opal has been used in the crowns of kings. The crown jewels of France and Hungary included fine specimens of opal. The crown of the Holy Roman Emperor contained an opal stone. Napoleon presented his Empress Josephine with the "Burning of Troy", a magnificent opal with brilliant red flashes.

Queen Victoria became a lover of opal and wore opals throughout her reign. Her friends and five daughters were presented with fine opals. Opal became highly sought-after because the Royal Court of England was regarded as the model for fashion around the world.

It was not long before famous designers such as Tiffany, Lalique and Cartier began using opal extensively in their jewellery.

In 1889 a new source of opal was discovered in far-off Australia producing the finest opal the world had ever seen. Australia's vibrant opal industry began with pioneering opal miners who worked (and sometimes died) in the remote Australian outback.

"... such resplendent lustre ..."

The Natural History of the World was written by Pliny the Elder in the First Century AD and was first published in English in 1601. Although Pliny suggests opal came from India, it did in fact originate in Hungary.

Pliny wrote: "For in the opal you shall see the burning fire of the carbuncle or ruby, the glorious purple of the amethyst, the green sea of the emerald and all glittering together, mixed after an incredible manner.

"Some opals carry such resplendent lustre with them that they are able to match the bravest and richest colours of painters: others represent the flaming fire of brimstone, yea and the bright blaze of burning oil."

Pliny also wrote that opals achieved the greatest price because of their rarity and their beauty. He writes of "Those who have written books of precious stone have given unto them the name and glorie of the greatest price."

Commenting on the value of opals, he indicates that, then as now, it was women who set the value of gems: "Thus far I have written of gems and precious stones which be esteemed principally and most rich, according to the decree set down and pronounced by our nice and costly dames."

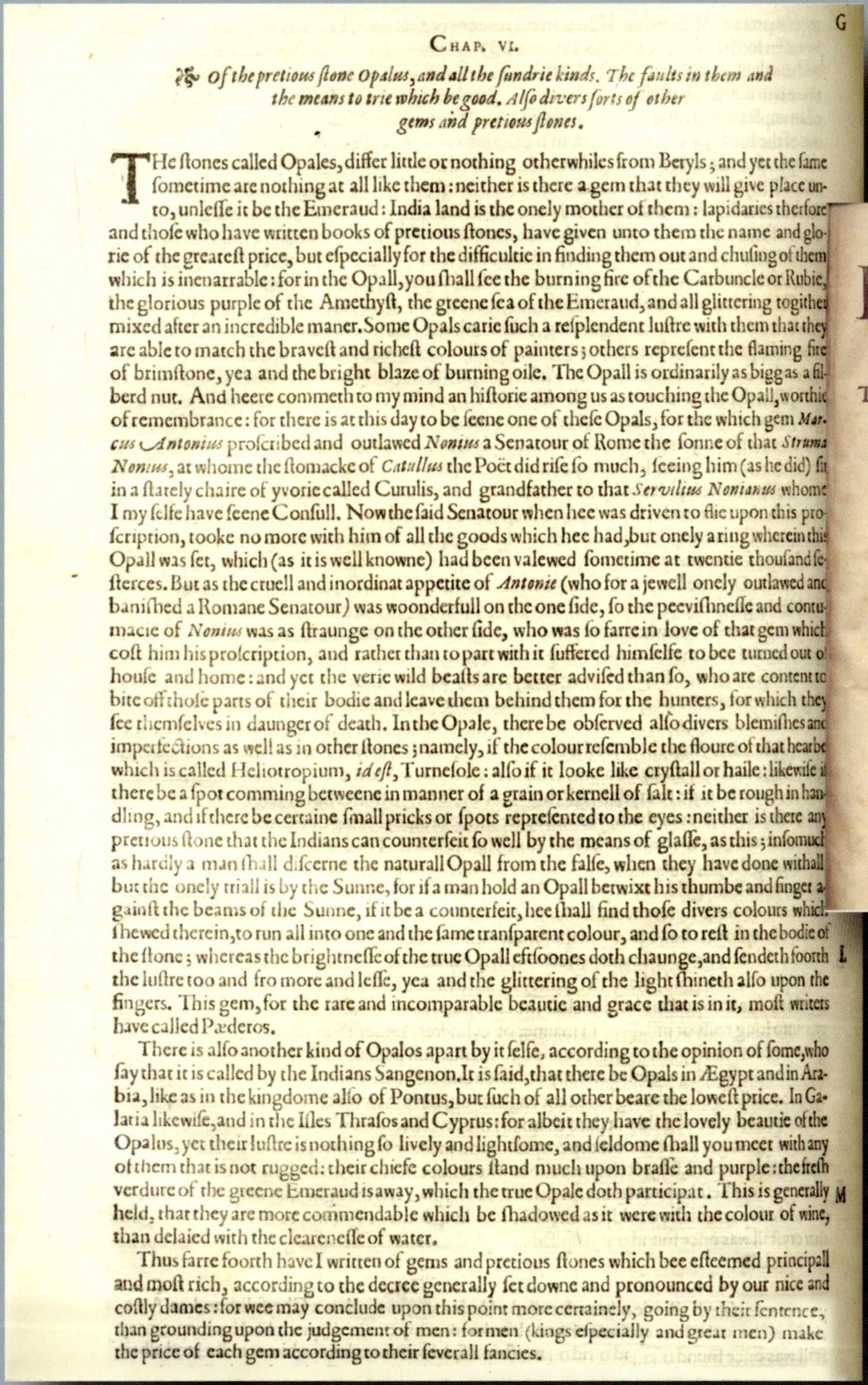

G

CHAP. VI.

Of the pretious ſtone Opalus, and all the ſundrie kinds. The faults in them and the means to trie which be good. Alſo divers ſorts of other gems and pretious ſtones.

THe ſtones called Opales, differ little or nothing otherwhiles from Beryls; and yet the ſame ſometime are nothing at all like them: neither is there a gem that they will give place unto, unleſſe it be the Emeraud: India land is the onely mother of them: lapidaries therfore and thoſe who have written books of pretious ſtones, have given unto them the name and glorie of the greateſt price, but eſpecially for the difficultie in finding them out and chuſing of them which is inenarrable: for in the Opall, you ſhall ſee the burning fire of the Carbuncle or Rubie, the glorious purple of the Amethyſt, the greene ſea of the Emeraud, and all glittering togither mixed after an incredible maner. Some Opals carie ſuch a reſplendent luſtre with them that they are able to match the braveſt and richeſt colours of painters; others repreſent the flaming fire of brimſtone, yea and the bright blaze of burning oile. The Opall is ordinarily as bigg as a filberd nut. And heere commeth to my mind an hiſtorie among us as touching the Opall, worthie of remembrance: for there is at this day to be ſeene one of theſe Opals, for the which gem *Marcus Antonius* proſcribed and outlawed *Nonius* a Senatour of Rome the ſonne of that *Struma Nonius*, at whome the ſtomacke of *Catullus* the Poët did riſe ſo much, ſeeing him (as he did) sit in a ſtately chaire of yvorie called Curulis, and grandfather to that *Servilius Nonianus* whome I my ſelfe have ſeene Conſull. Now the ſaid Senatour when hee was driven to flie upon this proſcription, tooke no more with him of all the goods which hee had, but onely a ring wherein this Opall was ſet, which (as it is well knowne) had been valewed ſometime at twentie thouſand ſeſterces. But as the cruell and inordinat appetite of *Antonie* (who for a jewell onely outlawed and baniſhed a Romane Senatour) was woonderfull on the one ſide, ſo the peeviſhneſſe and contumacie of *Nonius* was as ſtraunge on the other ſide, who was ſo farre in love of that gem which coſt him his proſcription, and rather than to part with it ſuffered himſelfe to bee turned out of houſe and home: and yet the verie wild beaſts are better adviſed than ſo, who are content to bite off thoſe parts of their bodie and leave them behind them for the hunters, for which they ſee themſelves in daunger of death. In the Opale, there be obſerved alſo divers blemiſhes and imperfections as well as in other ſtones; namely, if the colour reſemble the floure of that hearbe which is called Heliotropium, *id eſt*, Turneſole: alſo if it looke like cryſtall or haile: likewiſe if there be a ſpot comming betweene in manner of a grain or kernell of ſalt: if it be rough in handling, and if there be certaine ſmall pricks or ſpots repreſented to the eyes: neither is there any pretious ſtone that the Indians can counterfeit ſo well by the means of glaſſe, as this; inſomuch as hardly a man ſhall diſcerne the naturall Opall from the falſe, when they have done withall: but the onely triall is by the Sunne, for if a man hold an Opall betwixt his thumbe and finger againſt the beams of the Sunne, if it be a counterfeit, hee ſhall find thoſe divers colours which ſhewed therein, to run all into one and the ſame tranſparent colour, and ſo to reſt in the bodie of the ſtone; whereas the brightneſſe of the true Opall eftſoones doth chaunge, and ſendeth foorth the luſtre too and fro more and leſſe, yea and the glittering of the light ſhineth alſo upon the fingers. This gem, for the rare and incomparable beautie and grace that is in it, moſt writers have called Pæderos. L

There is alſo another kind of Opalos apart by it ſelfe, according to the opinion of ſome, who ſay that it is called by the Indians Sangenon. It is ſaid, that there be Opals in Ægypt and in Arabia, like as in the kingdome alſo of Pontus, but ſuch of all other beare the loweſt price. In Galatia likewiſe, and in the Iſles Thraſos and Cyprus: for albeit they have the lovely beautie of the Opalus, yet their luſtre is nothing ſo lively and lightſome, and ſeldome ſhall you meet with any of them that is not rugged: their chiefe colours ſtand much upon braſſe and purple: the freſh verdure of the greene Emeraud is away, which the true Opale doth participat. This is generally M held, that they are more commendable which be ſhadowed as it were with the colour of wine, than delaied with the cleareneſſe of water.

Thus farre foorth have I written of gems and pretious ſtones which bee eſteemed principall and moſt rich, according to the decree generally ſet downe and pronounced by our nice and coſtly dames: for wee may conclude upon this point more certainely, going by their ſentence, than grounding upon the judgement of men: for men (kings eſpecially and great men) make the price of each gem according to their ſeverall fancies.

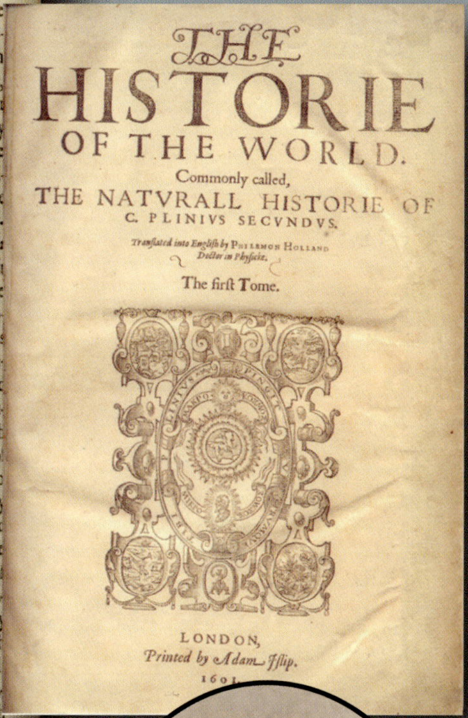

Some take the talismanic value of stones very seriously!
Photo from: "The Magic of Jewels & Charms" by George Frederick Kunz, 1913.

Shakespeare was attributed with the description of opal as "that miracle and queen of gems".

Opals Are Lucky

Throughout history opal has been regarded as the stone of good fortune.

Late in the 19th and early in the 20th century however, most probably for commercial reasons, there was a proliferation of damaging reports that opal was an unlucky stone. These were based upon a misinterpretation. There can be little doubt that much of the modern superstition regarding the supposed unlucky quality of the opal owes its origin to a mis-reading of Sir Walter Scott's novel. "Anne of Geierstein".

"The wonderful tale … contains nothing to indicate that Scott really meant to represent opal as unlucky."*

In Roman times the gem was carried as a good luck charm or talisman, as it was believed that an opal brought its owner good fortune. Today, in a number of cultures, opal is carried or worn as a good luck charm. Lucky opal, the rainbow gem, the stone of hope, the birthstone of October.

**The Curious Lore of Precious Stones, by George Frederick Kunz, 1913.*

The Opal in Australian History

Tullie Cornthwaite Wollaston, born in South Australia in 1863, braved life-threatening conditions to acquire opal from early miners in remote, desolate areas. It was Tullie Wollaston who introduced Australian opal to the rest of the world. He is considered the father of the opal industry.

In the early 1900s Wollaston introduced black opal overseas. At first the dealers in Hatton Garden, London, thought the opal must have been treated as they had never seen opals so bright and beautiful.

Wollaston established many markets in Europe and the United States during the early 1900s.

Australia now produces around 95 per cent of the world's opal supply and Australian opal is prized the world over, thanks to the pioneering efforts of Tullie Wollaston.

Tullie Wollaston from "Tullie Cornthwaite Wollaston" by D. P. T. Wollaston.

(Above left) Workings at the Three Mile Lightning Ridge. Circa 1903.

(Left) Sunday at the Three-Mile Black Opal field. Circa 1906.

Photos from "Opal: The Gem of the Never Never" by T. C. Wollaston.

Opal Formation

Across the world precious opal occurs in very few locations because its formation requires very special circumstances involving the coincidence of geological and possibly biological phenomena.

Scientific research and studies over the last two centuries have provided comprehensive detail about the formation of most gemstones. Unfortunately the genesis of opal is still not properly understood.

The results of thorough research will enable better targeted prospecting which could have a marked impact on the future supplies of opal and the livelihoods of thousands at the opal fields.

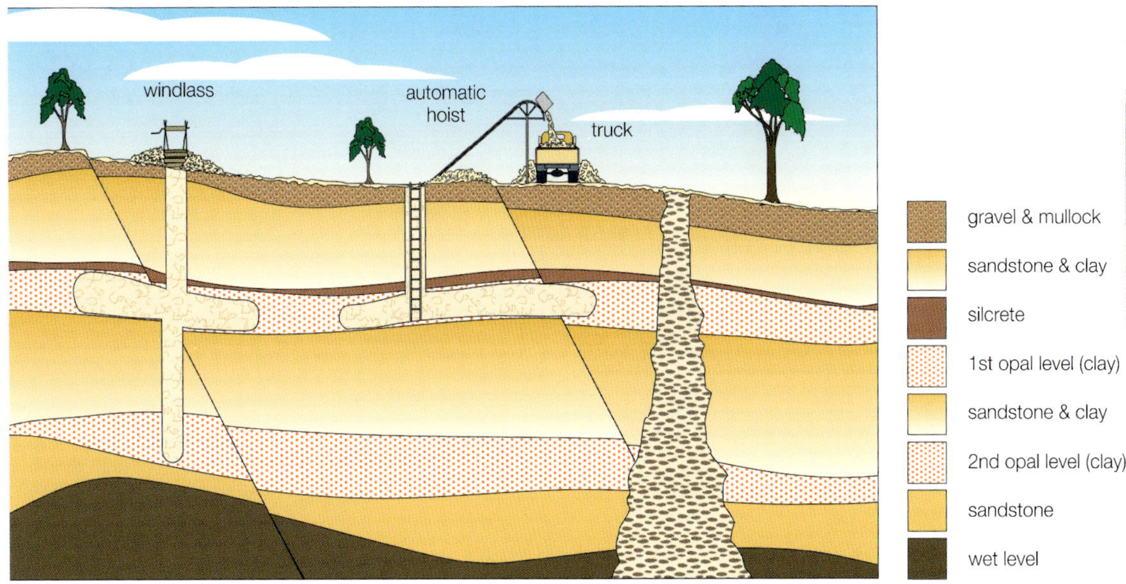

Slide (fault). *Vertical "blow" or Breccia zone.*

The Deep Weathering Model

This theory has been the oldest and most widely accepted explanation since the 1960s. During the mid-Tertiary period, around 30 million years ago, deep weathering caused chemical changes to the sediments releasing large quantities of soluble silica, which percolated downwards and along faults to an impermeable level. Here it precipitated within voids and cracks as common opal and on rare occasions it formed colourful precious opal.

The Syntectonic Model

Some scientists believe that opal formed via the silica-laden fluids moving through the layers of sub-surface strata under fault-generated hydraulic pressure and at temperatures greater than 100°C. The faults, caused by tectonic processes, create pathways for fluids super saturated by amorphous silica. This fluid, while under pressure, becomes entrapped. It cools, gels and hardens to form opal.

The Microbe Model

With the birth of the new millennium came a new theory from German scientist Dr. Hans-Jurgen Behr, who found evidence of rich aerobic soil bacteria and microbes conserved in opal. The bacteria would act as a "bio-reactor" resulting in the creation of enzymes and acids important in the formation of opal.

There is now evidence to suggest that opalisation in Australia's sedimentary deposits occurred at the same time as the sediments themselves were laid down. The presence of aerobic (oxygen requiring) microbial remains within the opal and the discovery of opalisation of soft tissue lends weight to the theory that the opalisation process occurred quickly during the Early Cretaceous period, over 100 million years ago.

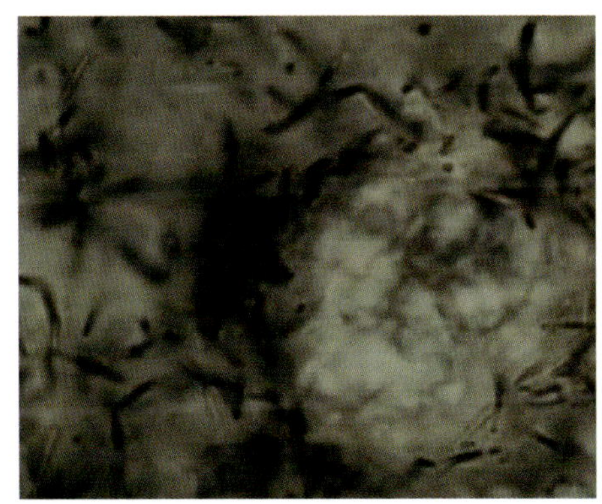

Microbes in opal (Photo: H. J. Behr).

This incredible stone continues to mystify the scientific world and is becoming increasingly elusive to miners and prospectors.

Opal Curiosities

Australian opal produces some remarkable geological specimens. This is largely due to the unusual set of geological circumstances during its formation. Opal specimens have been found which take the shape of other minerals, including calcite glendonites after Ikaite. Some specimens have ancient water encapsulated within them, while opals from all fields have been found to contain microbes. Occasionally opals display pictures such as animal shapes or landscapes.

Extraordinary collection items also include a hypsilophodontid femur which has tooth-like structures embedded in the bone, suggesting an attack from which the animal survived - as the bone has grown around the injury.

When "Nessie", a three-metre opalised pliosaur, was discovered (photo page 15), there was also evidence of its last supper; opalised fish vertebrae and a belemite cone (a squid-like animal) were found, together with gastroliths (stomach stones).

Opal sometimes arranges itself in shapes and colours resembling landscapes or animals. These are known as "picture stones". Here, a pair of magpie birds can be seen.

Fibrous silicified wood within opal potch. Lightning Ridge.

Possible borer in opalised wood. White Cliffs.

(Right) Opalised crinoid "sea lily". Coober Pedy.

Australian volcanic opal specimens from Tintenbar.

Australian volcanic opal specimen from Byron Bay.

Beautiful gem crystal in a geode formation.

Opal "pineapples" are a replacement structure called a glendonite, which is the name given to the calcite pseudomorph structure that forms after destabilisation of Ikaite at temperatures greater than six degrees Celsius. The opal did not replace the Ikaite itself but rather the secondary calcite - the opal could therefore be interpreted as a "tertiary replacement".

"The Duck" - twinned nobby with water and air inclusion. Lightning Ridge.

Opal enhydrite - ancient water and air trapped inside an opal.

Prehistoric Puzzle

(Top and above in close-up)
The hypsilophodontid femur from a cretaceous dinosaur, with two pathological lumps which appear to be teeth or spurs imbedded in the leg bone. The tooth- like structures conjure images of an attack from another creature, but the dinosaur recovered from the injury as the bone shows signs of healing around the teeth or spurs.

(Left) A model of a hypsilophodontid dinosaur.

Opal and the Dinosaurs

By strange coincidence and only on rare occasions, fossil bones from the Early Cretaceous period (115-100 million years) have been found which have been replaced by opal. In exceptional circumstances the opal can exhibit a brilliant play of colour.

Opalised remains from the "Age of the dinosaurs" have been important for understanding past environments and the evolution of the strange creatures that still exist in Australia today.

Opal module displaying marine snail in gem quality black opal.

(Below) Artist's impression of the Lightning Ridge area of inland Australia 100 million years ago.

(Above) Unique and priceless opalised upper jawbone from Kollikodon, a bizarre early cretaceous mammal.

"Nessie", a three meter long opalised pliosaur on display at the National Opal Collection in Sydney.

(Right) Opalised jawbones of the mammal Steropodon.

(Left) Opalised snails, Lightning Ridge.

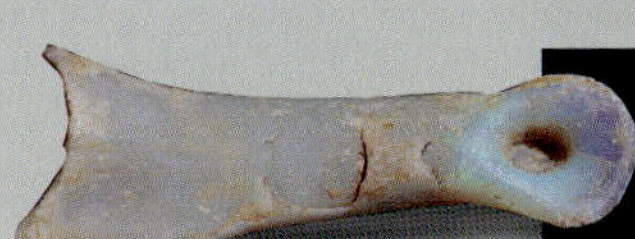

Opalised toe bone from a carnivorous dinosaur (theropod) , Lightning Ridge..

Pliosaur rib bone.

Opalised mussel, Coober Pedy.

(Below) Artist's impression of the pliosaur (left) and (right) another marine creature, an ichthyosaur.

Examples of Opalisation of Prehistoric Fossils

Many opalised fossils are highly significant to international science. The molecule-by-molecule replacement of the bone by opal often preserves the fine details of the internal anatomy.

Many fine examples of opalised fossil remains of dinosaurs, marine reptiles, fish, mammals and invertebrates have been found, providing a fascinating insight into Australia's prehistoric past.

The Rainbow Gem

Colours in Opal

Precious opal is composed of minute uniform spheres of silica, which are arranged together in an orderly three-dimensional grid. The spaces between these spheres contain silica in solution. However, in common opal, where there is no play of colour, the spheres are of varying sizes and are not uniformly stacked.

(Right) "The Path of Enlightenment" an exceptional opal necklace with 86 crystal opals.

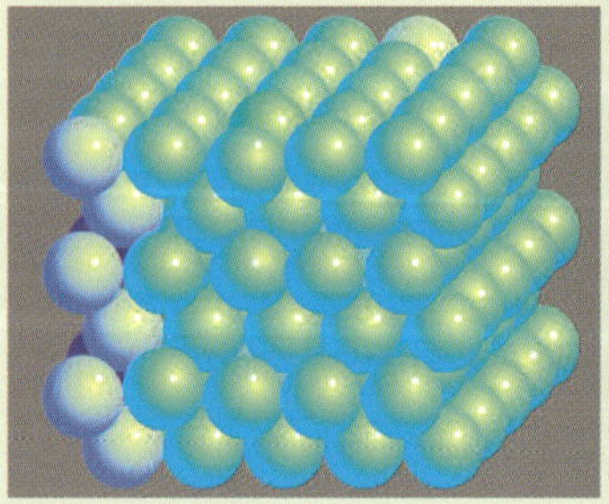

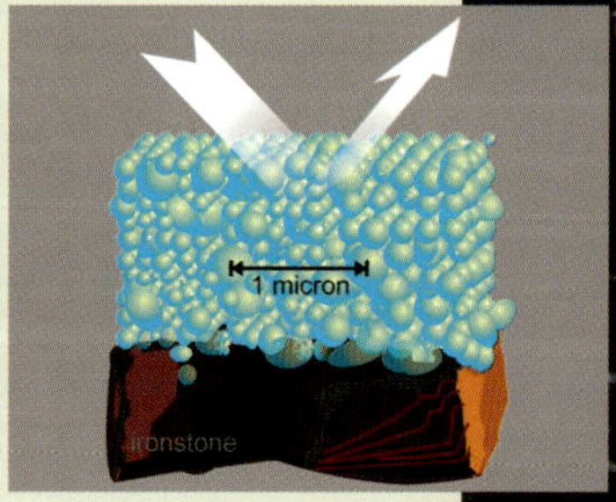

Precious opal: *minute spheres of silica produce diffraction of light into the colours of the rainbow.*

Diagrams courtesy of Opalinda.

Common opal: *the microspheres are of varying size and there is no orderly arrangement.*

White light passes through the transparent spheres directly, but when it reaches the silica in solution it is deflected at angles. These diffracted beams of light may show all the colours of the spectrum, or particular colours may predominate.

The colour emitted from opal is dependent on the size of the spheres, which determine the wavelengths. For instance, blue colours are evident where the spheres are smaller and, at the other end of the spectrum, orange and red will be evident where the spheres are larger. The intensity and brilliance of the colours are a result of the degree of uniformity of size and regularity of the grid.

Precious opal with orderly grid on the upper surface, diffracting white light into the colours of the spectrum.

Diagram courtesy of Opalinda.

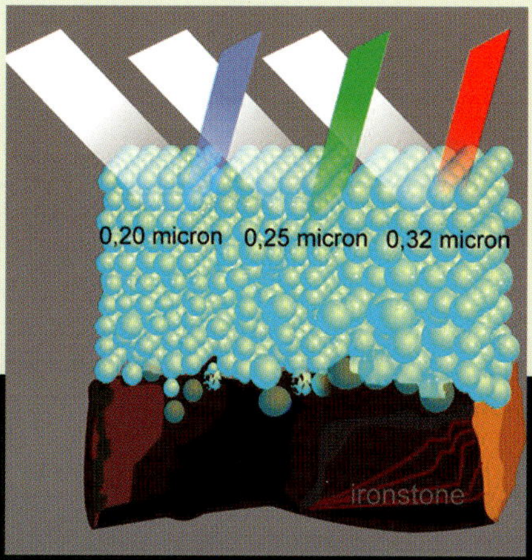

Types of Opal

Black (and Dark) Opal

Black opal is a generic term given to any opal which has a dark body colour when viewed from above. "Play of colour" within a dark or black body colour will enhance the brilliance of the colours. Some have a transparent crystal colour bar overlaying dark opal potch, giving the otherwise light opal a dark appearance. Even expensive black opals may have only a thin colour bar on black potch.

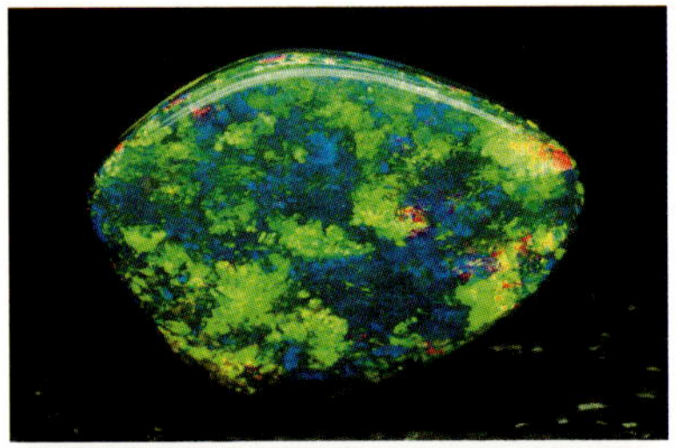

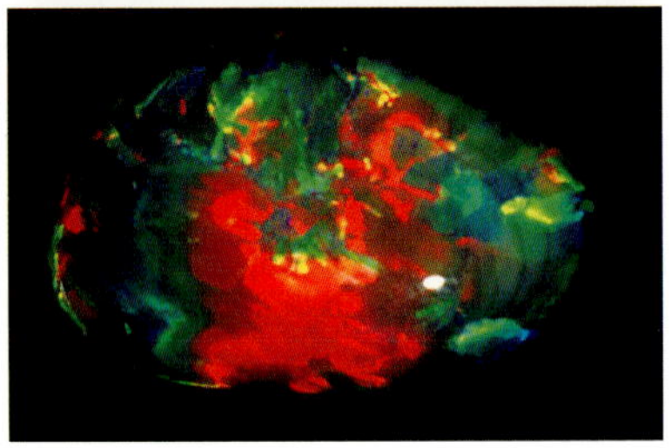

Gem quality black opals.

A rare matching pair of black harlequin opals.

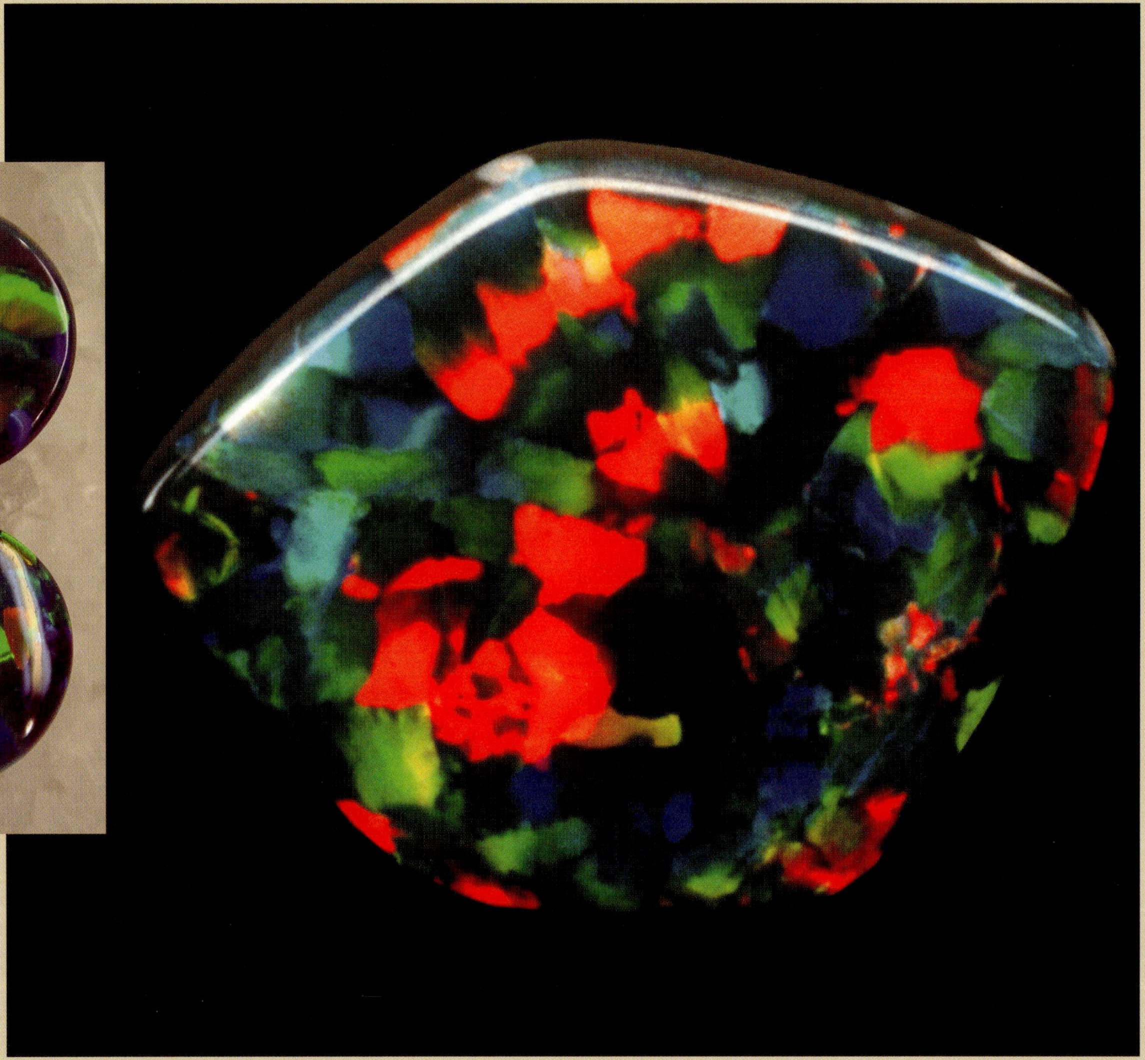

"The Millennium" black opal.

Black Opal
The body colour of this type of opal is dark grey to jet-black. Black opal is the rarest and most valuable opal variety. The example photographed (above) is "The Millennium", a truly magnificent and rare harlequin black opal weighing 25 carats.

Gem black opal.

Rough black opal, Lightning Ridge.
Photography by Maha Tannous, GIA.

Automatic hoist at the Three Mile Field, Lightning Ridge.

"Cody's Camp", Three-Mile Field, Lightning Ridge, used during the 1970s and now owned by the Lightning Ridge Historical Society.

Underground mine at Lightning Ridge.

Types of Opal

Boulder Opal

Boulder is a variety of precious opal that forms naturally within the host rock; often just a thin vein of precious opal is present. It mainly occurs in specific locations over a wide area of Western Queensland, formed as the in-fillings of cracks or voids, usually in ironstone boulders.

Boulder opal can be black or light depending on the appearance when viewing from the face of the opal. The popularity of this type has soared since the mid-1970s. Boulder opals are invariably cut incorporating the host brown ironstone. They are in high demand and can be extremely valuable. A variety of boulder opal also occurs in Andamooka, South Australia, where the host rock is quartzite. These are referred to as "painted ladies".

Photo: Alex Lane.

"The Galaxy", perhaps the finest boulder opal ever found. It weighs a staggering 520 grams or 2600 carats.

Boulder Opal
Fine grade boulder opals have the top surface covered with precious opal. Slight marks and protrusions of ironstone in the face of the stone are fairly typical, and can affect value. Boulder opal is usually cut to the contours of the opal vein creating a baroque wavy surface and they are often irregular in shape. The most expensive boulder opals are those with a dark body tone and displaying a full face of brilliant precious opal.

Boulder Matrix
Veins or patches of precious opal intimately mixed with the host rock are referred to as boulder matrix opal. This form is found mainly in Queensland. Matrix opal increases in value with brilliance and as the quantity of precious opal becomes a higher proportion of the face of the stone. Fine boulder matrix invariably has attractive banded ironstone with high polish.

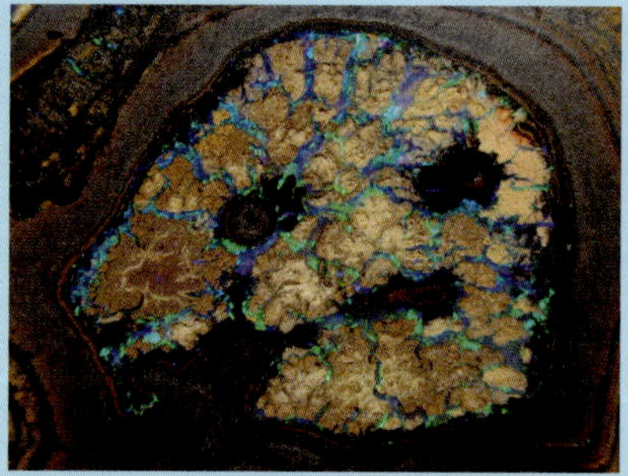

Boulder Splits
These are formed where a single piece of rough is split along the opal vein to form a matching pair. If the cleaved top surface is mostly precious opal of brilliant colour, this variety can be very valuable.

Boulder opal rough.

Opal Nuts
These are commonly referred to as "Yowah Nuts" although they are mined at both Yowah and Koroit in southwestern Queensland. They derive their name from their nut-like appearance. The opal occurs in kernels inside an ironstone concretion. The ironstone in the opal nuts is usually an attractive red colour and extremely hard, therefore it will polish to a high gloss. These are sought after as collection pieces and for jewellery.

Types of Opal

Light Opal

Light Opal is commonly referred to as "white", although this term should be used only where the body colour is milky or white. It should not be used to describe the entire light opal family. Light opal makes up the bulk of precious opal and comes from all fields but the majority is found in South Australia.

Brilliant light opal can be extremely valuable. The transparent material (crystal) variety commands the highest price.

Digging for opal in rock deep beneath the earth in Coober Pedy.

Rough Light Crystal Opal
A sliver of rough crystal opal prior to cutting and polishing.

Milky (White) Opal
This is the most common type of precious opal, and is translucent to opaque with a creamy appearance, dominating the diffracted colours.

Regular Light Opal
The base colour ranges from white to medium grey. The diffracted colours are brighter than in milky opal and often the stone has a degree of transparency.

The amazing underground community of Coober Pedy in South Australia, where miners seek relief from the hot climate in dugout homes. Summer temperatures often exceed 50C.

Grey Opal
This opal is almost always opaque to translucent ranging from light grey to pewter colour. Opal darker than this is classed as dark or black opal.

Jelly Opal
A transparent type, which has an appearance rather like gelatine. The colours are indistinct and the colour flecks are not concentrated in the stone.

Light Crystal Opal
This group embraces those opals which are transparent enough to view the colour patches below the surface or within the gem. Crystal opals can exhibit brilliant colour play.

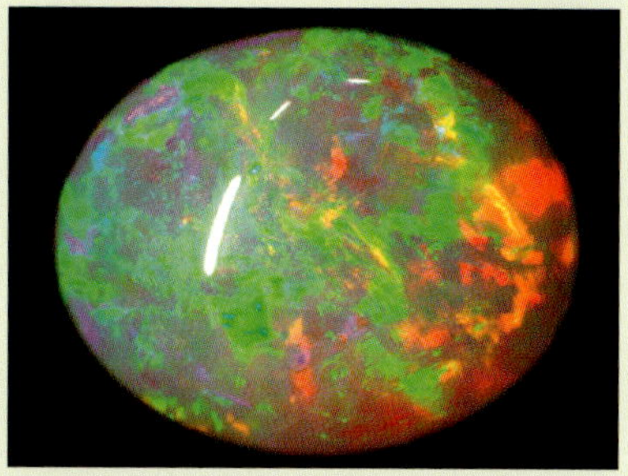

A "ripper" being used in Mintabie to expose opal at the opal level. Photo by Andrew Cody.

Types of Opal

Composite Opal

Composite opals comprise natural opal manually attached to another substance. Triplets and doublets are a combination of natural opals and artificially attached veneers. It is important not to submerse composite opals in water or detergents as this could cause deterioration of the glues used to cement the lamination.

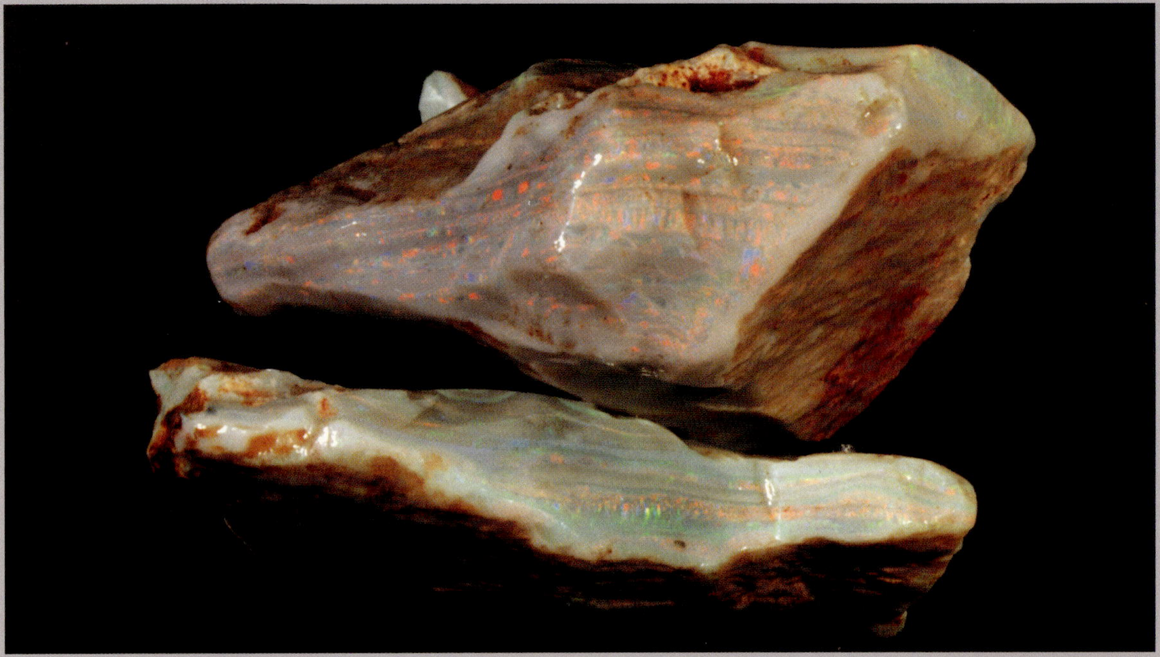

Seam rough opal recovered from The Olympic Field, Coober Pedy. Total weight 30oz (almost one kilogram).

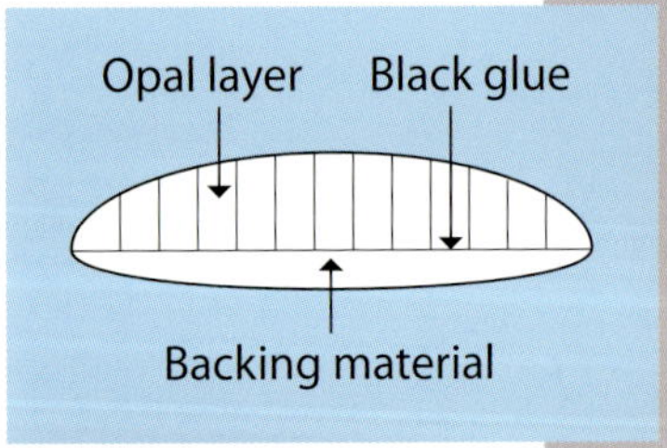

Doublet Opal
A doublet opal is a layer of precious opal bonded to a black backing, simulating black opal. Doublets are usually more expensive than triplets because more opal is used.

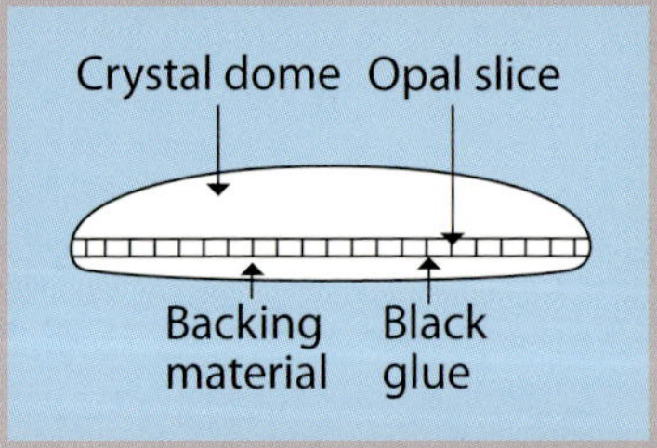

Triplet Opal
A triplet opal is a thin veneer of precious opal bonded to a black backing with a protective crystal dome glued to the top.

(Left) Perhaps the largest piece of full gem colour rough opal ever found. Lambina opal fields, 1990.

(Below) The same opal parcel showing strong colour bars.

(Above) Grading a parcel of light opal after cutting.

Valuing Opal

There is no governing body which has set the prices for opal; they are the result of what the international market is prepared to pay in order to obtain the gem. As with any other commodity, opal is subject to the laws of supply and demand. The supply of opal is never guaranteed - the luck of the miner, variable costs, new discoveries, floods and weather conditions - all affect the supply of the stone. Demand relates to the enthusiasm of the customer for opal, and may be influenced by the world economy, fashion and market trends.

Unprecedented demand over the last three decades for fine quality opal has caused prices to soar. Prices of fine quality black opals rival those of diamonds.

The attractiveness and appeal of a particular opal are the factors that are important at the final point of sale. This is discussed on the following pages.

Selection of exceptional black opal.

Factors Determining Value

1. Type of Opal

It is important to determine the type of opal. For instance, a dark boulder opal may produce a similar colour display to a black opal. However, the black opal will command a higher price. The main types of Australian opal are depicted below. Black, boulder, boulder matrix and light opal are all natural forms. Treated Andamooka matrix opal is subjected to solutions of sugar and sulphuric acid to darken the body tone and enhance the colour. Composite opals, such as doublets and triplets, contain slices of opal cemented to other materials.

Black

Boulder

Boulder Matrix

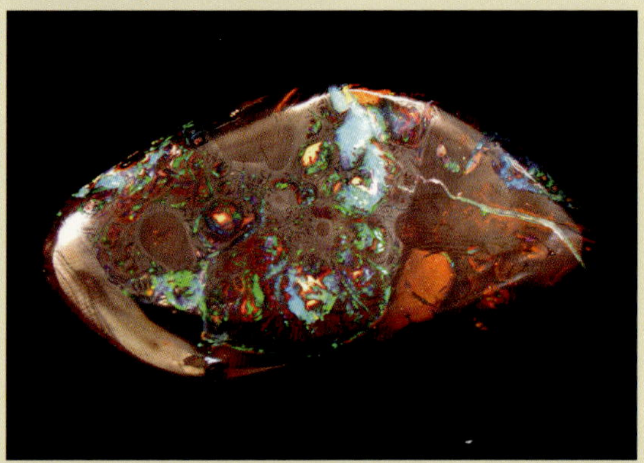

Light

Treated Matrix (Andamooka type)

Composite (doublet)

Factors Determining Value

2. Brilliance

The degree of brightness is of paramount importance. An opal cannot be valuable without this attribute. A gem that glows but is average in all other respects will still command a high price whereas a dull stone, with a very good pattern and range of colours, may bring only an average price. The stone on the bottom right has the highest degree of brilliance.

Factors Determining Value

3. Body tone

The degree of darkness in the body tone, in regard to black or boulder opal, is taken into consideration when assessing the value. A black body colour is more valuable than a grey or light body tone, when all other factors are equal.

N1

N2

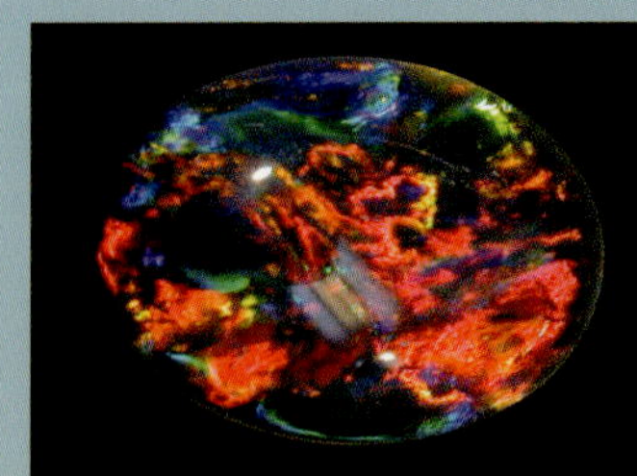

N3

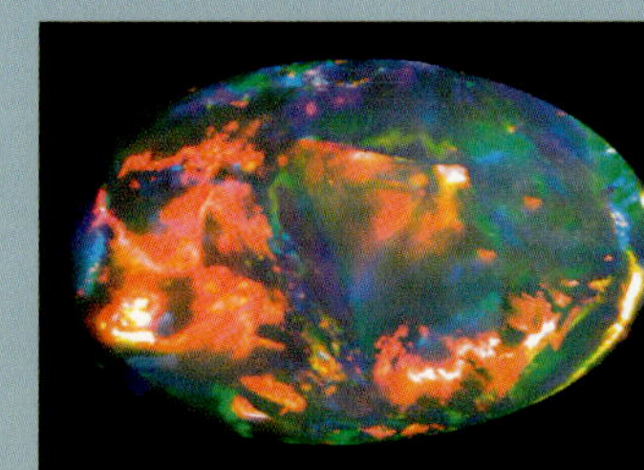

N4

A group of opals illustrating the body tone variations. These variations can be measured against tone patches (below each image) taken from a chart prepared by the Geological Society of America for ascertaining body tones .

N5

N6

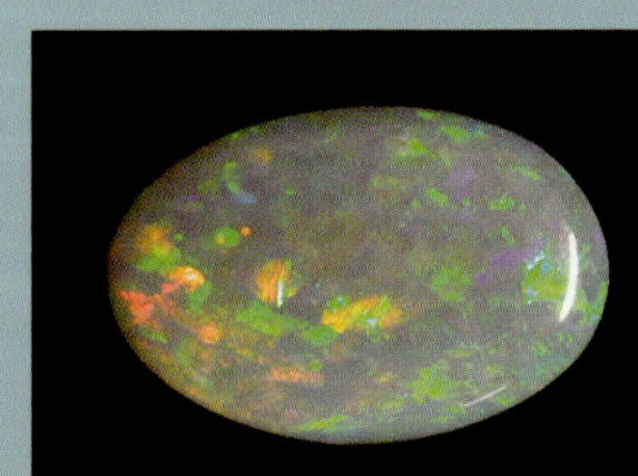

N7

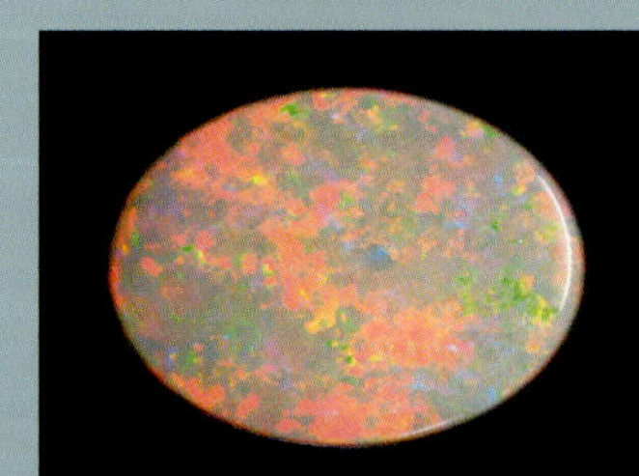

N8

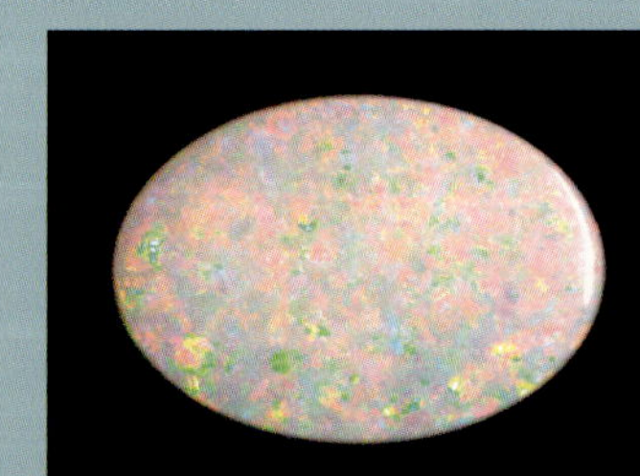

N9

Factors Determining Value

4. Transparency

Light opal is more desirable when it has a degree of transparency and if lively, brilliant colours are present in crystal opals, they are most highly prized.

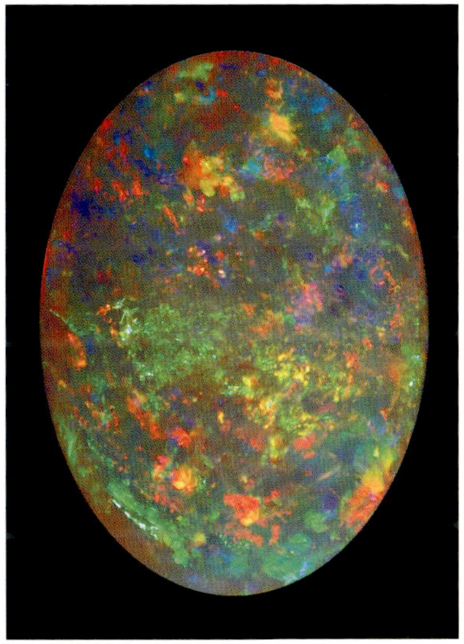

Crystal rough.

Butterfly wings.

Crystal on black background.

Same crystal opal indicating transparency.

Black crystal opal.

Factors Determining Value

Interesting patterns of the diffracted colours are rare and, when combined with brilliance, have an enormous impact on the value. Pinfire and small type patterns are less desirable than broad patterns or large flashes. Distinct patterns such as rolling flash, straw pattern, Chinese writing, ribbon and harlequin are highly sought after.

Rare starburst pattern displayed in the "Supernova" black opal (10.75 carats).

5. Pattern

Pinfire.

Peacock tail.

Block.

Jigsaw.

Harlequin.

Flagstone.

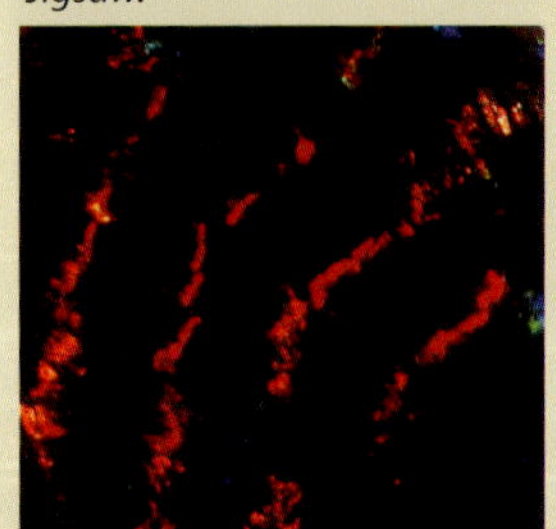

Ribbon.

Peacock or fan harlequin.

Straw.

Broad flash.

Chinese writing pattern.

Stripes.

The Rolling Flash has a pattern much like a cat's eye - a line of colour which rolls across the face of the stone.

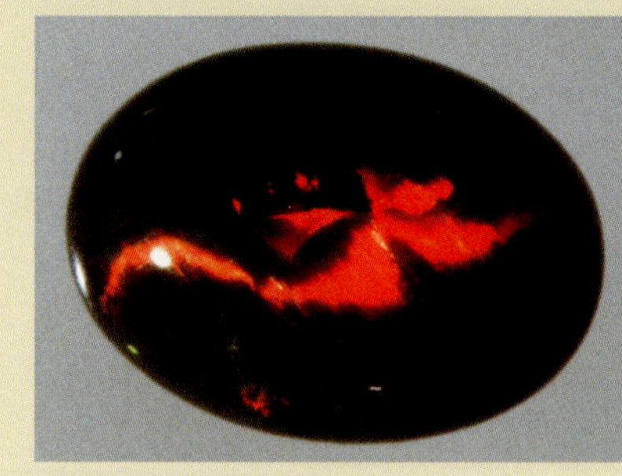

Rolling Flash - view from one direction.

Rolling Flash - same stone as above from the opposite direction.

A Broad Flash pattern displayed in a stone known as the "Opera House".

Factors Determining Value

Orange multi colour gem block opal set in platinum with diamonds.

Photo courtesy of Jean Francois Albert.

6. Dominant Diffracted Colours

Great care needs to be taken when applying this factor, remembering that type and brilliance override all other factors. A blue/green brilliant stone will usually be more valuable than a dull red stone. The dominant colour has value in this order: red, orange, yellow, green, blue, indigo and violet. The red and orange colours are the more valued. This rule is more particular to black opals. Strong red colour play in a dark body-toned black opal is the most valuable.

Blue.

Blue green.

Green blue.

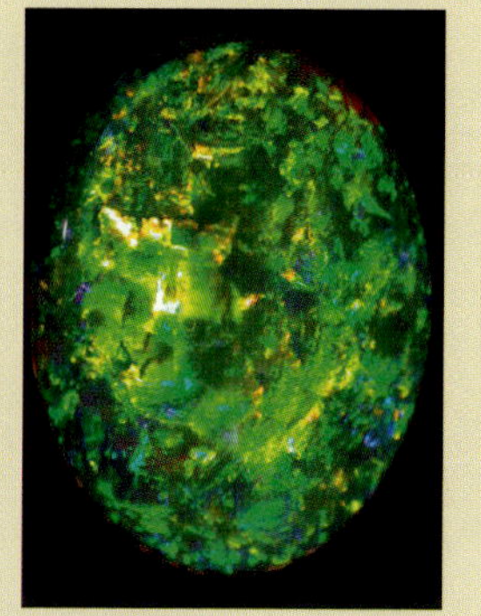

Green yellow.

Orange blue green.

Orange multi colour.

Red multi colour.

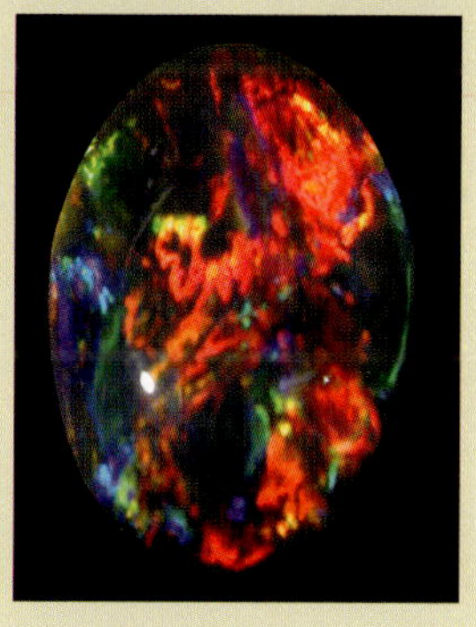

Crimson multi colour.

Red multi colour.

Factors Determining Value

7. Thickness of the colour bar

Black opals and black boulder opals with a very thin colour bar are frequently worth many thousands of dollars per carat, although they would be worth more if they had exactly the same appearance with a thicker colour bar. Thickness of the bar enables a domed surface to be cut from the rough, thereby increasing value.

High dome.

Medium dome.

Low dome.

Factors Determining Value

Butterfly wing.

8. Shapes

Precious opal is usually cut into oval, pear or round shapes. Occasionally other shapes are cut including marquis, rectangle and heart. Non-standard shapes include freeform and baroque. Shapes which will produce difficulty for the jewellery setter will often decrease the value of a stone.

Oval.

Ovals.

Free shape - 4-sided.

Oval.

Free shape - pear.

Free shape - triangle.

Drop shape - bullet.

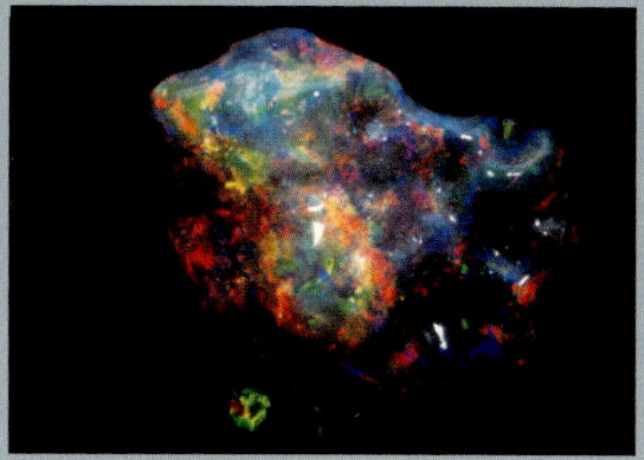

Baroque.

Free shape - kite.

Care of Opal

Australian opal is found in sedimentary rocks aged 115-100 million years. The majority of opal is quite hard and will scratch glass.

However there are a few areas at some fields that produce unstable material with a higher water content, which can crack or craze. It is important to buy your opal from reputable traders who guarantee their stones.

Solid opal

Australian opal has a typical hardness of 6 to 7 on Mohs' scale of hardness from 1 to 10. (Pearl = 3.5, glass = 5 , tanzanite = 6.5, emerald = 7.75, sapphire = 9, diamond = 10)

Opal is quite resilient, but (as is the case with any valuable jewellery) care should be taken to avoid situations where the opal could come in contact with abrasives or be subject to impact. Thick stones are more resilient than thin stones.

If you intend to store an opal for a long period in a safe or safety deposit, it is recommended the item be placed in a sealed container or bag with water or oil. Safes are a de-humidified environment and are unnaturally drying. Follow these simple rules and your opal will give you a lifetime of pleasure and become a treasured heirloom.

Triplets, doublets and inlay

It is very important to protect doublets and inlay opals from impact. On some occasions the opal is just a thin veneer, which is likely to crack if dropped on a hard surface. As triplets, doublets and inlay are laminates the glue could deteriorate if submersed in water and particularly detergents. Water or detergent does *not* affect solid opals.

Japanese inspired jewellery by Marianne Hunter

Photos by George Post.

(Right) Kendal Schuler, 2008 Ambassador for the Australian Opal Industry, wearing a spectacular black opal set weighing 540 carats.

The Opal Fields

The major outcrops of opal in Australia occur along the shoreline of what was once an ancient inland sea (known as the Eromanga Sea). The important mining areas in Australian opal production over the last century have been White Cliffs, the Queensland fields, Lightning Ridge, Grawin, Glengarry, Coober Pedy, Andamooka, Mintabie and Lambina.

White Cliffs

Commercial mining commenced in the late 1880s at the White Cliffs Field in New South Wales. The opal mined here excited the world and the town soon developed; by the 1920s it had a population of more than 3,000 people. White Cliffs produced fine quality seam material in such volumes as the world had not previously seen. Sadly, the main run was worked out by 1930 and the field all but died out.

Queensland fields

The Queensland fields are spread across 1,000 kilometres (600 miles) and produce almost all the world's supply of boulder opal.

This opal is produced from remote locations in the vicinity of Yowah,

White Cliffs, NSW.

Queensland Fields.

Eromanga, Quilpie, Jundah, Opalton and Winton. Almost all the opal is won by open cut mining. Large and small ironstone boulders are split in the search for that elusive opal vein.

Lightning Ridge

This field was discovered in 1905 by kangaroo shooters. "The Ridge" produces almost all of Australia's sensational black opals and, despite a recent decline in production, it is still the largest producer of opal by value. (Lightning Ridge opal is typically found as nodules or "nobbies".)

In the late 1980s the Coorcoran field 30 km to the west of the township produced stunning black nobby opal in great quantities.

The population soared to over 8,000 people. At this time opal sold for record prices sometimes at more per carat than good quality diamond. Outlying fields include The Grawin, Glengarry, Sheep Yards and Mulga Rush, which predominantly produce dark and black seam opal.

Andamooka

Andamooka is in the harsh desert and located near the edge of Lake Torrens (a saltpan over 165k or 100 miles in length). The field was discovered in 1935 and is famous for producing brilliant crystal seam opal.

Andamooka reached its peak in the 1960s when the population reached 3,000. Unfortunately this field is worked out, with very little commercial production since

Lightning Ridge, NSW.

Andamooka, South Australia.

Coober Pedy, South Australia.

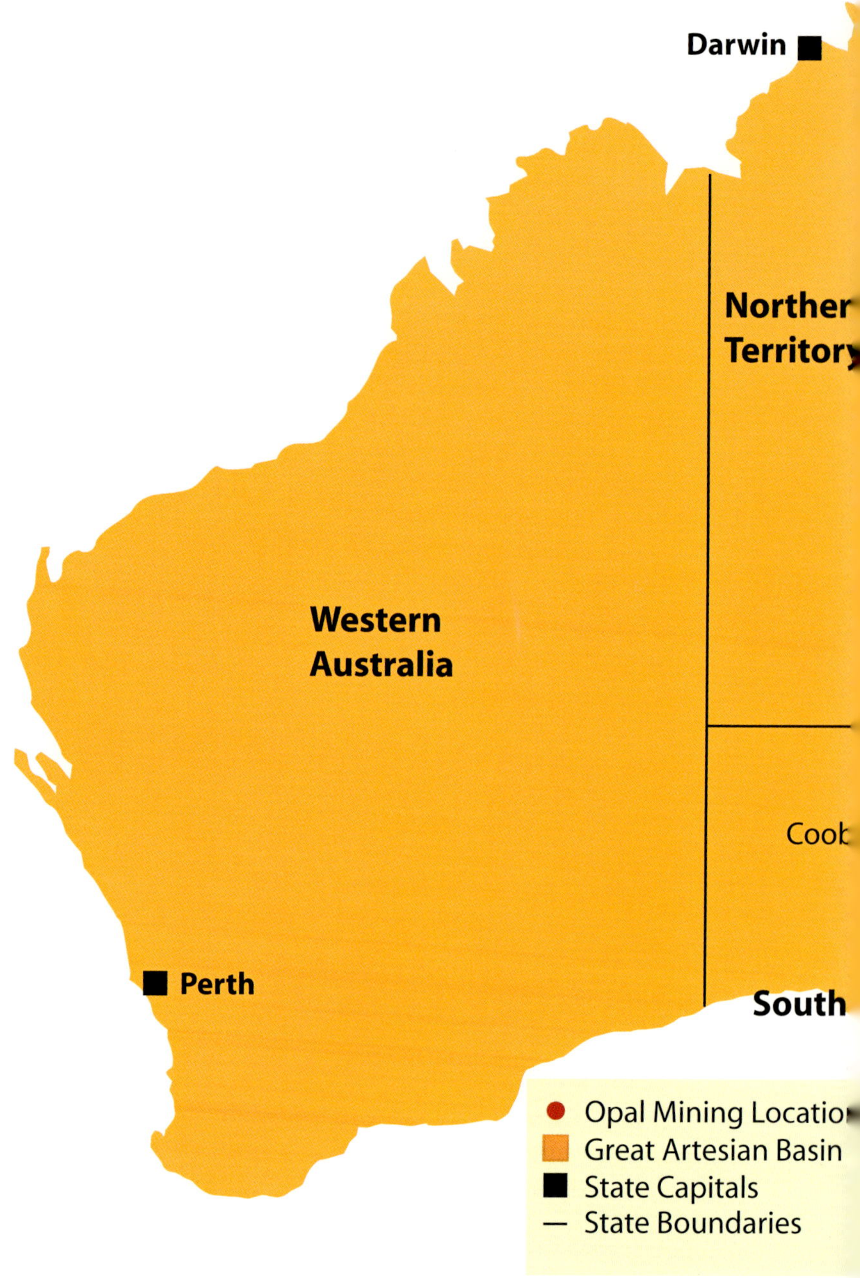

the 1970s. Today much of Andamooka's population is employed at the nearby uranium mines of Roxby Downs.

Coober Pedy

The bulk of the world's light seam opal has been mined at Coober Pedy. Mining extends over an area 50 kilometres long, running approximately north/south.

The town reached its peak in the 1970s when enormous quantities of opal were being produced. It attracted many migrants from all over the world, seeking to make their fortune. Illegal gambling joints proliferated, some with dozens of tables where tens of thousands of dollars changed hands.

Unbearable summer temperatures have resulted in most of the population living underground in dugouts. By 2008 Coober Pedy's production had fallen to 15% of its heyday but this dusty, hot outback town remains home to more than 4,000 people representing at least 50 nations.

Mintabie

The fortunes of Mintabie came and went within a decade. In 1978 a group from Coober Pedy found good quality black opal. Within a few years the population grew to 1,500. Large powerful equipment was needed to rip the hard sandstone. More than 75 large bulldozers, together with many scrapers and excavators, quickly mined out this very rich outcrop. Mintabie produced fine quality light and black seam opal. This material was extremely hard; in some cases it is able to scratch agate. Today there is only very limited mining activity.

Lambina

This field steadily grew from 1989, when production from neighbouring Mintabie was in sharp decline. In 2002 Lambina was the only shining light of the South Australian opal fields, with over 300 miners working with bulldozers and excavators. Sadly, by 2005, this field too was also in decline. The material is typically light and dark seam opal.

Lambina, South Australia.

Mintabie, South Australia.

The National Opal Collection Directors

ANDREW CODY

Building on a consuming interest in science and gemstones from early childhood, Andrew established a wholesale and cutting business in 1971 which later expanded to include exporting to Europe, Japan and the USA.

In 1999, he became an Associate Research Officer of the Australian Museum. He has served the industry in a number of official capacities including President of the Australian Gem Industry Association and Foundation Member and Chairman of the Australian Jewellery and Gemstone Industry Council. In 2007 he became World President of the International Colored Gemstone Association (ICA).

Andrew's first publication, "Australian Precious Opal - a Guidebook for Professionals" is used extensively worldwide by the industry. It has been published in English and Japanese.

DAMIEN CODY

After completing a Diploma in Business Studies, Damien carved a highly successful career in the insurance and financial services sectors where he held senior executive positions with a number of leading international organisations.

His passion eventually led him to the opal industry where he and his brother are directors of a leading Australian opal exporting company which also operates retail opal stores and museums in Australia. These companies have won Government Export and Tourism Awards for excellence in these fields.

Damien has served the industry in a number of capacities including as International Colored Gemstone Association Ambassador to Australia.

MAX LANE

Max has a degree in Gemmology and has a long history in the gemstone industry in Australia. He has served as Federal President of the Australian Opal and Gemstone Association (AOGIA) and Chairman of the Australian Jewellery and Gemstone Industry Council.

A qualified pilot, he regularly flies to the opal fields of Australia where he has long been a familiar figure.

Max founded his business in 1971 and built it into one of Australia's largest opal companies. He has been an active participant in the Australian opal industry for more than 40 years and is renowned globally as a leading exporter of boulder opal.

Photography

RUDY WEBER

Rudy's interest in gemstones began in the 1970s, under the influence of his gem collector father, and coinciding with a love of photography. After studying diamond grading in the UK, he returned to Sydney to set up a gem laboratory and photographic library business.

His fascination for the opal is evident in his extensive photographic library, capturing the essence of the stone, the outback mining locations and the characters who mine it. His businesses have received prestigious Australian Achiever Awards for excellence, while his work can be found in countless publications, gemmological literature, posters, educational books and displays around the world.